Tons of Trucks!

GARBAGE TRUCKS

By Martin Harasymiw

Gareth Stevens
PUBLISHING

Please visit our website, www.garethstevens.com. For a free color catalog of all our high-quality books, call toll free 1-800-542-2595 or fax 1-877-542-2596.

Library of Congress Cataloging-in-Publication Data
Names: Harasymiw, Martin, 1978- author.
Title: Garbage trucks / Martin Harasymiw.
Description: New York : Gareth Stevens Publishing, [2022] | Series: Tons of trucks! | Includes index. | Audience: Grades K-1.
Identifiers: LCCN 2020024877 (print) | LCCN 2020024878 (ebook) | ISBN 9781538262832 (library binding) | ISBN 9781538262818 (paperback) | ISBN 9781538262825 (set) | ISBN 9781538262849 (ebook)
Subjects: LCSH: Refuse collection vehicles–Juvenile literature. | Refuse collection–Juvenile literature.
Classification: LCC TD794 .H37 2022 (print) | LCC TD794 (ebook) | DDC 628.4/420284–dc23
LC record available at https://lccn.loc.gov/2020024877
LC ebook record available at https://lccn.loc.gov/2020024878

First Edition

Published in 2022 by
Gareth Stevens Publishing
111 East 14th Street, Suite 349
New York, NY 10003

Editor: Therese Shea
Designer: Deanna Paternostro

Photo credits: Cover Alex Stemmer/Shutterstock.com; p. 5 Sergiy Kuzmin/Shutterstock.com; pp. 7, 19 Andyqwe/iStock/Getty Images Plus/Getty Images; pp. 9, 24 (cab) Kzenon/Shutterstock.com; pp. 11, 24 (hopper) kozmoat98/E+/Getty Images; p. 13 Paul Vasarhelyi/Shutterstock.com; pp. 15, 24 (fork) SunChan/E+/Getty Images; p. 17 PeopleImages/iStock/Getty Images Plus/Getty Images; p. 21 Dalibor Danilovic/Shutterstock.com; p. 23 Dmitry Kalinovsky/Shutterstock.com.

Printed in the United States of America

CPSIA compliance information: Batch #CSGS22: For further information contact Gareth Stevens, New York, New York at 1-800-542-2595.

Contents

This is my toy truck.
It's a garbage truck!

This is a real one!
It picks up trash.

The driver
sits in the front.
The front is the cab.

The trash goes in
the back.
The back is the hopper.

Some trucks have an arm.
The arm picks up cans.

STOP
ARRÊT

Some trucks
have a fork.
It picks up big bins.

Truck parts push down trash. They make space.

The trucks take
the trash away.

The back lifts up.
The trash falls out.

These trucks help!

Words to Know

cab

fork

hopper

Index